Hummingbird and Ostrich

A Bird Book for Kids™

By Novare Lawrence

Nada Bindu Publishing Co.

The contents of this book previously appeared in the digital-only editions *Hummingbird: A Bird Book for Kids* and *Ostrich: A Bird Book for Kids*.

First Print Edition – December 2014

ISBN-10: 1633070077
ISBN-13: 978-1-63307-007-3

Published by:
Nada Bindu Publishing Co.
Carson City, NV 89703
Website: www.nadabindupublishing.com
Email: inquiries@nadabindupublishing.com

CONTENTS

Hummingbird

Hummingbirds are amazing. They are among the smallest, fastest and most beautiful birds in the world. Hummingbirds live in many different countries throughout North America, Central America, South America and the Caribbean. Over fifteen different types of hummingbirds can be found in the United States. Their feathers reflect bright colors like red, green, orange, yellow and blue and they sparkle in the sunlight when the light shines on them just right.

A Green Violetear hummingbird

A hummingbird's body is only around two to three inches (5-7.6 cm) long, not counting its long beak or tail feathers. Compare the tiny hummingbird to the biggest bird you'll find today, the ostrich: most adult hummingbirds weigh less than a U.S. nickel or about 0.14 ounces (4 g), but an adult male ostrich can weigh over 300 pounds (136 kg) which is 4800 ounces (136078 g). The smallest hummingbird is the Bee Hummingbird found in Cuba – it weighs less than a single U.S. penny.

The smallest and the biggest birds

Hummingbirds start out very small indeed. Hummingbird eggs are about the size of a jellybean and weigh only about .03 ounces (1 g). In fact, most hummingbird nests are smaller than a single chicken egg. Ostrich eggs are the largest bird eggs in the world and weigh between 2-1/2 and 5 pounds (1.1 - 2.3 kg).

Ostrich Egg

Hummingbird Nest and Egg

"Robin" Eggs

Chicken Eggs

You're very lucky if you ever find a hummingbird nest in your yard because they are so small and hard to see. Of course, you should never move or disturb a hummingbird nest. Let the mother hummingbird take care of the baby hummingbirds so that they grow up healthy on their own – she is very smart and knows just what to do and how to care for her young. A hummingbird mother will sit on her eggs until they hatch. After they hatch, she will spend most of the day finding food to bring back to the chicks.

Two hummingbird chicks waiting for Mom to bring food

A hummingbird builds her nest using small grasses, feathers, leaves, twigs, even spider webs. With such a small nest, most hummingbirds lay only two tiny eggs. After they hatch out of their tiny eggs, the new baby hummingbirds grow quickly and fill up the nest. After about three weeks of feeding by their mother, the chicks begin to exercise their wings until they are strong enough to fly away. Hummingbird chicks may fly with their mother for a while to learn where the best places to find food are but quickly go off on their own after that.

Hummingbird chicks ready to leave the nest

Hummingbirds fly fast even though they are so small. Most hummingbirds fly around their territory at about 30 miles per hour (49 km/hr) during normal daily activities, like searching for food and playing with other hummingbirds. But a hummingbird can also dive through the air at over 50 miles per hour (80 km/hr) if it has to chase or flee from another hummingbird, which makes it very hard to see them.

A hummingbird flying fast is pretty blurry

Although their small size makes it seem like they can disappear in a flash, hummingbirds are not the fastest birds around. In the air, the fastest bird is the peregrine falcon. When hunting and diving, peregrine falcons can reach top speeds over 200 miles per hour (322 kph). On the ground, big ostriches can run over 60 miles per hour (97 kph) for short distances so even they can travel faster than a hummingbird.

Peregrine
Falcon

Ostriches running

Fastest birds in the air and on land

Hummingbirds can fly long distances. Even though we think of hummingbirds as small, fast flyers, we don't often realize that many hummingbirds migrate with the seasons. The Rufous hummingbird has been known to fly all the way from Alaska to Mexico. Ruby-throated hummingbirds can fly from the east coast of the United States down as far south as Panama in Central America. That trip includes a flight of more than 500 miles (805 km) over water without stopping. Hummingbirds may migrate to find more food or due to weather as the seasons change but in most climates hummingbirds can usually find food and mates year round without migrating.

Hummingbirds and bees love nectar

It is true that because of their small size, hummingbirds seem to be even faster than they really are. They can disappear from your sight very quickly, zipping away in a straight line or making a quick change in direction that most other birds can't do. You almost can't even see their wings. Both their small size and their special wings allow them to fly in ways that other birds can't. A hummingbird's wings beat about 70 times per second but they can beat up to 200 times per second in order to fly extra fast. When their wings move that fast, the hummingbird's heart may have to beat up to 1000 beats per minute.

A Broad-tailed hummingbird

One of the special flying abilities that hummingbirds have is the ability to stop very quickly. To stop quickly or change direction, a hummingbird's tail feathers act as a fast braking system. After flying quickly from one spot to another, a hummingbird will flare out his or her tail feathers to form a bowl that captures the air to slow down instantly. It's like cupping your hand into the wind and feeling the air resistance instead of opening your fingers and letting the air flow through.

A Hummingbird's tail feathers help it stop quickly

But the real secret to a hummingbird's wings is what gives hummingbirds that special ability to hover in mid-air. Most birds can only flap their wings up and down as they lay flat when flying, but a hummingbird can flap its wings at any angle to its body, very fast forward and backward, in a pattern that looks like a figure-8. This pushes the air in the right direction allowing them to go fast, change direction, and fly backwards or to hover.

Hummingbirds can hover and fly "standing up"

Other birds try to hover but it takes a lot of effort. Some hunting birds like falcons and hawks can fly in one spot as they track a small animal on the ground but they can't do it for very long. And many types of seabirds and hunting birds can also stay almost motionless in the sky by turning their body and wings into the wind or using the thermal currents rising up off of hills. With only small adjustments in their wings, they can stay effortlessly in one place without flapping at all but this is actually gliding and not hovering. Hummingbirds don't glide at all – their wings are always in motion when they fly.

A Hawk gliding through the air

Hummingbirds do very practical things while hovering. They hover as they feed from flowers and they hover to watch other hummingbirds and visitors coming into their territory. Hummingbirds also eat small flying insects right out of the air by opening their beaks wide and catching them in between. With such special wings, the hummingbird can also fly and hover in a standing up position while other birds can only fly as if they are lying flat on their stomach.

Hummingbirds feeding and visiting

Hummingbird beaks are very long for such a small bird. Some hummingbird beaks are straight and some are curved. Most hummingbird beaks are about one-third of their entire body length but there is one hummingbird, the Sword-Billed hummingbird in South America, which has a beak that's even longer than its body. Hummingbird beaks evolved to match the types of flowers where they can find the nectar that they need to feed on. In addition to a long beak, the hummingbird has another special item: a very long tongue.

A Purple-throated hummingbird shows off the tip of his tongue

Hummingbirds have a special diet. They eat very small insects like gnats, mosquitoes and other flying bugs but their favorite food is sugary sweet flower nectar. In order to get to the nectar at the center of a flower, hummingbirds use their long beak and their long tongue. A hummingbird will hover at a group of flowers and patiently poke its long beak into each and every single flower. As they do this their long tongue dips into the small sticky droplets of nectar and pulls them back to swallow and enjoy.

A hummingbird feeding at lavender blossoms

Hummingbirds are very smart. For their small size, hummingbirds have more of their entire body weight in their brain than any other bird. Hummingbirds can remember the best flowers to visit, both in their home territory and when they migrate. They even seem to remember how long it takes a flower to refill its nectar between visits and whether the nectar has enough sugar to give them the energy they need.

Tasting the nectar

You can help to attract hummingbirds to your yard by duplicating the nectar that most flowers produce. It's easy to do and will allow you to see hummingbirds feeding up close. First, you need a hummingbird feeder. Hummingbird feeders are usually red with yellow drinking holes to attract the hummingbirds and may hold one to two cups of sugar water (1/4 – 1/2 liter). Some have little perches where the hummingbird can sit or the hummingbird may just hover while feeding.

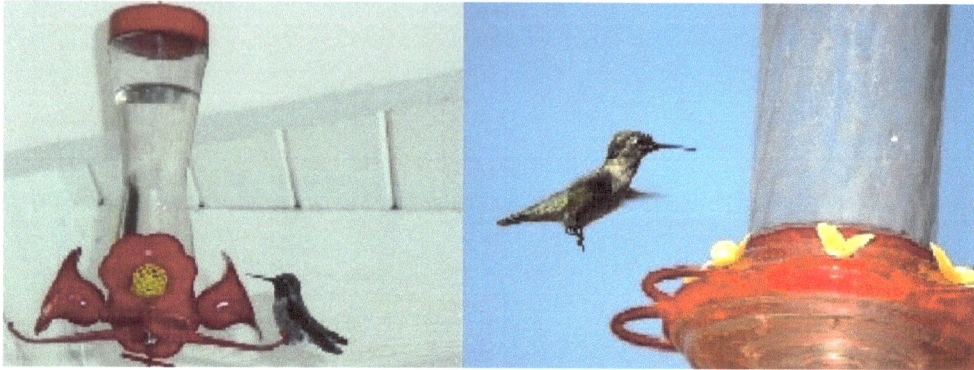

Hummingbirds love full hummingbird feeders

To make your nectar, use 1 part white sugar to 4 parts water. For example, to make one cup of sugar water, measure 1/4-cup of white sugar into a measuring cup and add warm or hot water to just over the 1-cup mark. Stir the water well to make sure that the sugar dissolves completely. When you make your sugar water, do not add red food coloring – the red and yellow colors of the feeder will attract the hummingbirds.

It is important to rinse out the feeder before using and to let your sugar water cool first before you pour it into the feeder. After you fill it, get an adult to help hang the feeder outside where it won't be bothered by any animals or accidently disturbed by people. The hummingbirds will find it within a day or two and you'll be able to watch them hover and feed on your sugar water.

Filling up with sugar water

Many people who put up a hummingbird feeder in their yard report that hummingbirds not only know who refills the feeder but that they will also come and watch that person through the window if the feeder has gone empty, as if to say, "Hey, where's the sugar water?" Just remember to refill your feeder when it is empty or at least every two weeks so that the sugar water stays fresh. Rinse out the feeder well before refilling it with fresh sugar water and you won't have any unhappy hummingbirds.

It's fun to have hummingbirds visit your yard every day. They will visit your feeder and any flowers in your yard. They may even play in your sprinklers to take a bath. They will get used to seeing you and will often fly right up to you to check you out, especially if you're wearing red. Hummingbirds can be quite playful and are a joy to watch.

Two Coronet hummingbirds visiting

Sometimes you may hear their singsong chirping, especially from the male hummingbirds as they mark their territory and tell other males to stay away. You might see a male hummingbird chasing another male out of the backyard while making sharp chirping sounds all the way. It's also possible to see a male and female hummingbird hovering together, facing each other while slowly spiraling upward as they get to know each other. Usually, it is the male hummingbird that has the brighter colors. This is because female hummingbirds need to blend in with the leaves and branches where they build their nest, keeping them and their eggs safe.

A male and female hummingbird hovering together

Hummingbirds eat and drink a lot for their size because their hearts and wings beat so fast. In order to slow down and rest at night, a hummingbird finds a safe branch on which to rest. Its claws lock on and the hummingbird uses its special ability to cool its body down almost as if it is hibernating. This resting condition is called a torpor. In the morning, as the air gets warmer with the sunrise, the hummingbird slowly wakes up, raising its body temperature. After relaxing its claws and doing a bit of stretching, it flies off again for a busy, busy day.

Flying uses up a lot of energy

And finally, to answer the age-old question, no, hummingbirds don't actually hum. It's the very fast vibration of their wings that makes the "humming" noise. You can often tell different hummingbirds apart just by the sound of their "hum" before you even see them. It's just another of the amazing things to like about hummingbirds.

Hummingbirds, in all the activities of their lives, remind us that even if you are small, you too can do amazing things.

Ostrich

Ostriches are the biggest, tallest and heaviest birds in the world. Some might even say that they are the funniest looking birds because of their long bare neck, long bare legs, and their big egg-shaped, heavily feathered body in between. You can tell the boys from the girls because male ostriches have black feathers while the female's feathers are a lighter sandy brown color.

A male and female ostrich on the plain

Unlike most other birds, ostriches don't fly. They are big birds! An adult male ostrich can weigh more than 300 pounds (136 kg) when fully grown and reach up to 9 feet (2.7 m) tall. The heaviest bird that can fly is another African bird called a Kori Bustard and even then, averaging around 46 pounds (21 kg) and 3 feet tall (.9 m), they only fly if they have to and only for short distances. So to get from place to place, ostriches walk or run. And ostriches with their long bare legs can run very well.

Three big ostriches running and a much smaller Kori Bustard who prefers walking to flying

Ostriches can run 40 miles per hour (64 kph) or more for short distances and around 30 miles an hour (48 kph) for longer distances. In Africa, where ostriches live, there is sometimes a very good reason to run fast. Africa is also home to some very good hunting animals, like lions, hyenas and especially cheetah. The cheetah can run between 45 to 60 miles per hour (72-96 kph) when chasing down its prey.

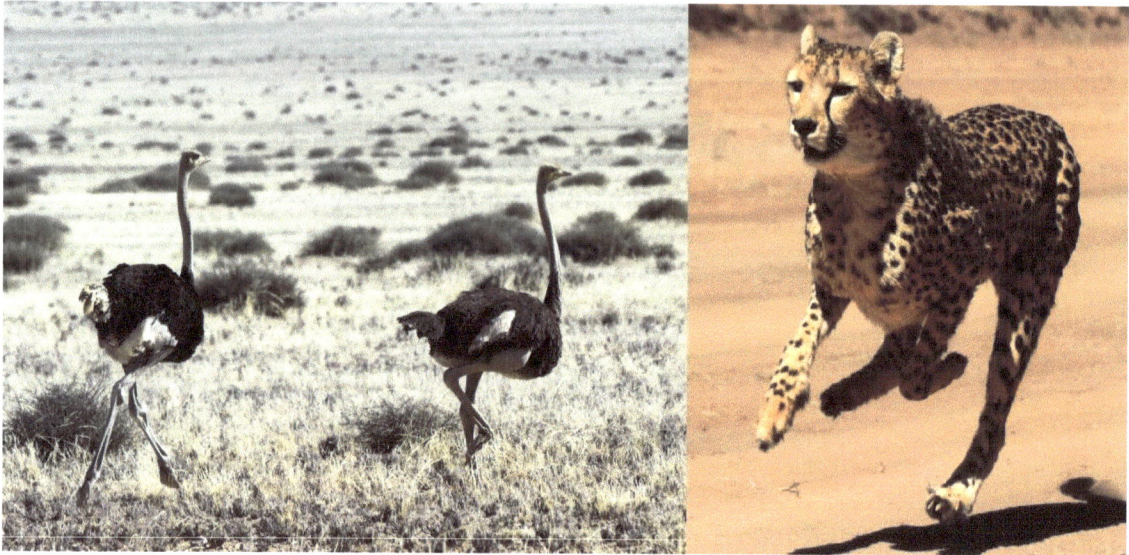

Ostriches keeping watch for hunters like this running Cheetah

Ostriches in the wild will often travel alone or with just one or two others, especially in winter. In summer and during the breeding season, ostriches like to band together and travel as a group. When predators are around, ostriches may travel with other animals on the plains such as zebra and gazelle, two animals that also run well. Having the extra company increases the security of the group, especially for the young ostriches.

Female ostrich with gazelles and zebras

Ostriches have some other advantages when it comes to staying safe. One of the reasons they run so fast is that their legs can cover up to 12 feet (3.6 m) with just a single stride. Their specialized feet help them out. Unlike other birds, the ostrich has only two toes on each foot. There is one big toe in front and a smaller one on the side used for balance. The front toe is long with a big claw that is used for gripping the ground as it runs. The claw is up to 4 inches (10 cm) long and is also used for defense. An ostrich will kick at any animal that threatens it.

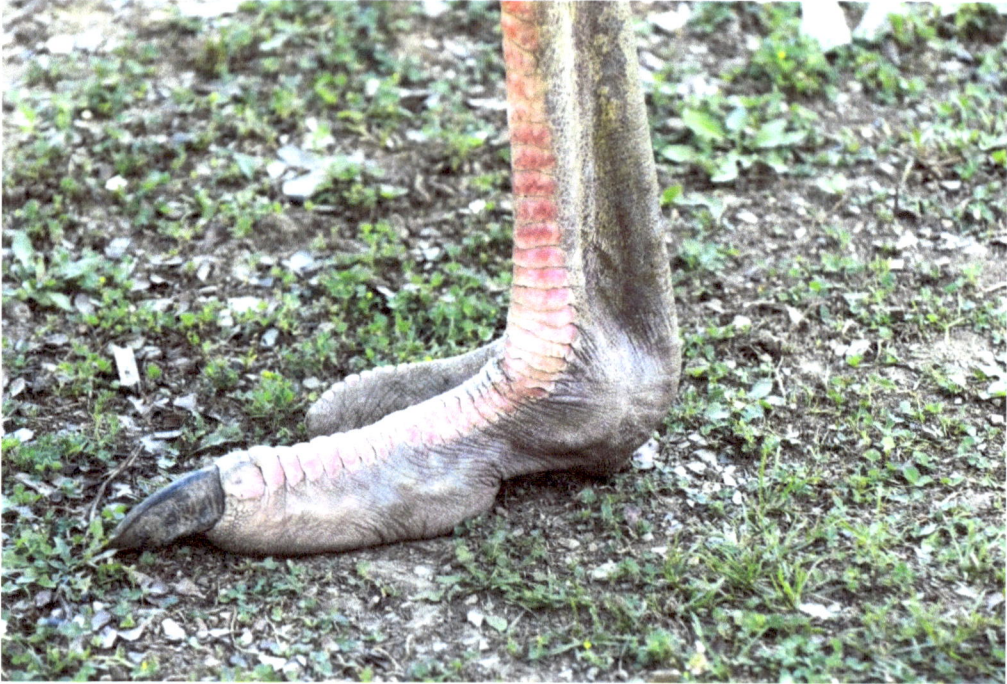

Ostrich feet have a big toe with a big claw

Ostriches also have the largest eyes of any land animals. Adults, with their great height, use those big eyes to see far into the distance and to look over shrubs, grasses and rocks where predators might hide. This means that they can see better than the predators that walk on four legs close to the ground. Their long neck also makes it easy for them to look around in every direction with little effort. It's not easy to sneak up on an ostrich.

Ostriches have big eyes with long eyelashes to protect them from the sun

In addition to the fact that ostriches are big and fast, if confronted, they can also spread and flap their big wings to look even larger and more threatening. When competing for leadership with other males within their group, male ostriches use their long necks and their large beaks to hit each other. This is also effective at intimidating any other animals that might be bothering them.

Male ostriches competing

Wings are very important to ostriches even though they don't fly. In order to run fast across uneven ground, an ostrich has to use its wings to help keep its balance and to change direction more easily. Ostrich wings are covered in feathers that are especially soft because they lack the stiff barbs that normally give feathers their strength. Flying birds need to have strong, stiff feathers in order to fly.

A soft ostrich feather compared to a rigid eagle feather

Big wings full of soft feathers are just perfect for ostriches. Their feathers help to keep their body temperature just right. When it's over 110°F (43° C) out on the open plains, their feathers help to protect their body from the harsh sun and keep them cool. At night when the temperature can drop from the high point of the day down to only 10 to 20° F (-12 to -6° C), ostriches sit on the ground and use their wings to cover their legs and neck in order to stay warm as they sleep.

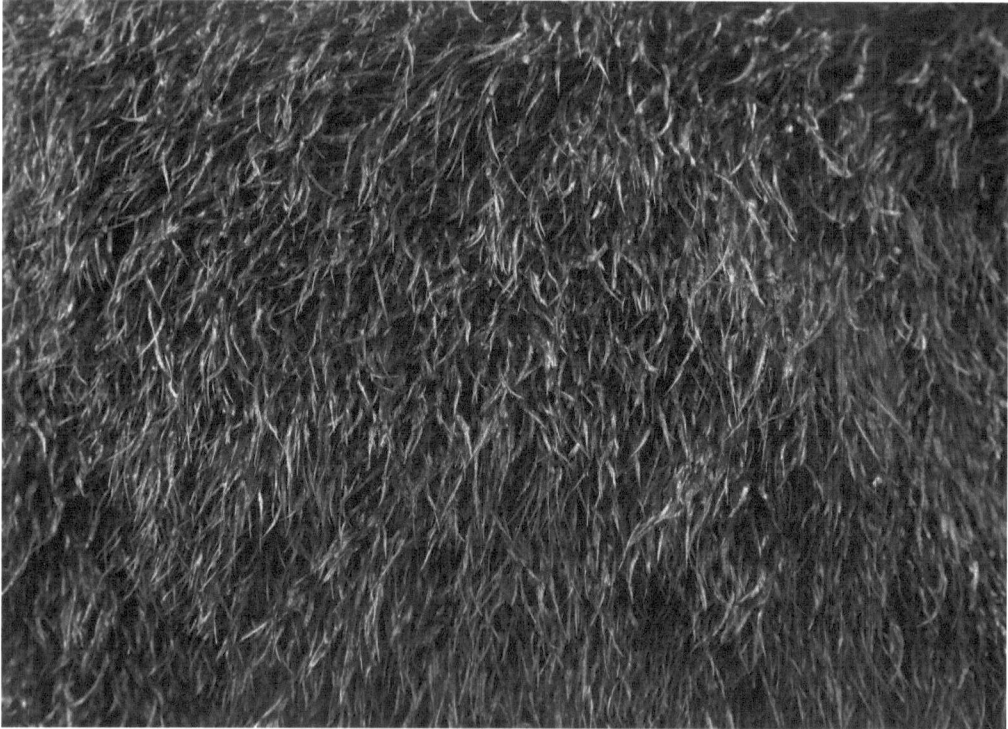

Thick layers of feathers protect the ostrich's pale white skin

The male ostrich uses his wings for another very special purpose. He dances to attract a female ostrich. In his dance, the male bends his legs at the knees and bobs up and down. He shakes first one wing, then the other and finally he shakes all his tail feathers. Then he slowly walks closer to the female with his wings held outward and his head and tail feathers held high. The male ostrich, with his black outer feathers and white under-feathers must look like a fancy tuxedo-dressed dancer as he circles around the female.

A male ostrich dances with his female partner

Feathers and wings help protect an ostrich in another way too. It may seem kind of funny, because they are so big, but an ostrich can hide just by sitting down. It works like this: if an ostrich traveling alone sees a possibly dangerous animal in the distance, it promptly sits down on the ground and covers itself with its wings. The potential predator, looking through the heat waves rising up from the ground, will see shrubs, grasses, rocks and an unmoving black or sandy brown lump that looks nothing like an animal. The predator usually ignores the lump but if it comes closer anyway, the ostrich can always jump up and run away. By sitting and waiting rather than running, the ostrich conserves its energy.

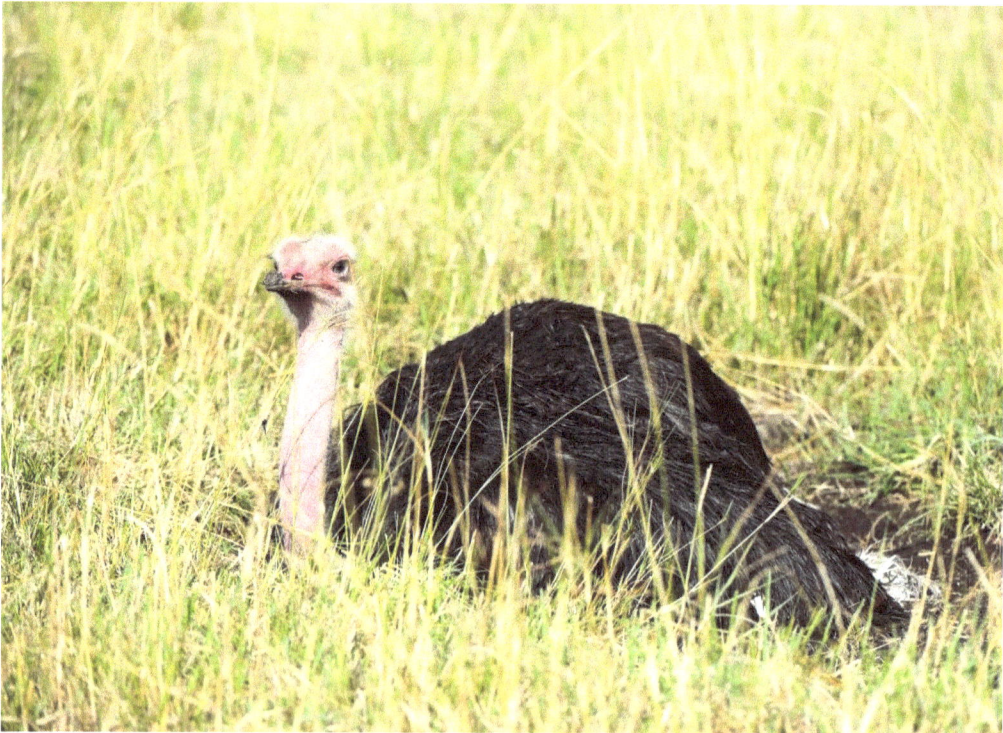

A male ostrich trying to look like a rock

One thing that ostriches don't do if they see a potential predator is bury their heads in the ground. Because ostriches are very tall and eat grasses and other plants close to the ground, it's not uncommon to see an ostrich with its head stretching down to graze. Seen from a distance, it might look as if its head is in the ground because it's the same color as its neck, but the ostrich is definitely not trying to bury it and hide.

A female ostrich grazing on grasses

Ostriches sit on the ground for another very important reason. That's where they make their nests. While many types of birds nest on the ground, an ostrich nest is very big. That's because ostriches like to travel in groups during breeding season. The females all lay their eggs in a single shared nest and take turns sitting on the eggs. A single ostrich egg can weigh up to 3 pounds (1.4 kg) – no other bird or reptile lays a bigger egg. With up to 20 eggs in a single nest, an ostrich nest can be over 9 feet (2.7 m) in diameter and 1 to 2 feet (0.3 – 0.6 m) deep.

An ostrich nest and one ostrich egg

One ostrich egg is twenty times heavier than a chicken egg and has about the same volume (the amount of space inside) as 24 chicken eggs. If you were to hard-boil an ostrich egg, it would take an hour and a half; a chicken egg only takes 12 minutes or so. The shell of an ostrich egg is pretty thick too so that it can support the weight of an adult ostrich sitting on top of it.

Ostrich egg compared to chicken eggs

Both male and female ostriches share when it comes to taking care of the eggs. Females usually sit on the eggs during the day while the males keep watch and then the males sit on them at night. This protects the eggs from the heat and cold. Sitting on the eggs also uses the same protection strategy of a single ostrich hiding in plain sight. Female ostriches have brown feathers that blend in with the ground and dry grasses during the day while the male with their black feathers blends perfectly into the dark night of the African plains. The other ostriches that remain standing nearby also provide security for the nest.

A female ostrich sitting on the eggs

Ostrich eggs hatch after around 40 days. All the adult ostriches in the group watch after the chicks, teach them what to eat and help to keep them safe. Ostrich chicks grow up pretty fast. They can grow by almost a foot (30 cm) taller every month during the first six months. And they are pretty fast too.

An ostrich chick just hatching from an egg while two chicks check out their home

By the time they are only one month old, ostrich chicks can already run up to 35 miles an hour (56 kph) for short distances. That is even faster than another bird that has a great reputation for running fast: the roadrunner. Roadrunners live in the desert areas of the southwestern U.S and in Mexico but adult roadrunners can only reach a top speed of about 20 miles per hour (32 kph) on the ground. Unlike ostriches, roadrunners can also fly if they have to.

Ostrich chicks are faster on the ground than a Roadrunner

Ostriches have an interesting diet. First of all, ostriches are able to survive on the hot African plains because they are able to go without drinking water for a long time, just like a camel. In fact, the scientific name for the ostrich species is *camelus*. Ostriches usually get most of their water from the plants they eat but they don't mind drinking it when it's available. They will even bathe in it if they find a large enough watering hole. If there is no water available, ostriches take dust baths by sitting on the ground and using their wings to scoop dirt up onto their body.

Ostriches taking a bath

As for eating, ostriches eat something very special that helps them digest everything else that they eat. They eat small rocks and even sand. Mammals, like cats and dogs and people, have teeth which we use to grind down our food before swallowing. This makes it easier for our stomach to digest and extract the nutritional ingredients and vitamins that our bodies need. Ostriches don't have teeth so they need small pebbles and sand in their stomachs to grind away at the food that they swallow whole.

An ostrich interrupted from its meal for a photo

For food, ostriches are mostly grazers, eating grasses, shrubs and other plants, as well as seeds and fruit. They may also eat insects, lizards or rodents if available. When an ostrich eats, it first collects food in the back of its throat and then, when it has enough, it swallows the whole ball of food. That way it's certain to go all the way down its long throat into its stomach.

Food travels a long way from an ostrich's mouth to its stomach

Wild ostriches are native only to Africa although they do live in zoos and farms all around the world. Like chickens and turkeys, some ostriches are raised for food. Ostriches can live up to 40 years in the wild and even up to 60 years in captivity where they receive a consistent, healthy diet. Whether they are in the wild or on a farm, ostriches are comfortable living together in groups. There is usually one male is in charge and his primary female partner leads all the other females.

Ostriches on a farm

There is a close relative to the African Ostrich that is often mistaken for one. This is the Australian Emu, sometimes called the Australian Ostrich. Emus look similar to their African relatives, eat a similar diet and practice similar habits, but emus are not as big and tall and so are considered to be only the second largest bird in the world.

The Australian Emu is smaller, lighter in color and its feathers cover more than half its neck.

If you ever have the chance to visit an ostrich at a zoo or farm or if you ever go to Africa, don't go up to an ostrich and try to pet it unless there is a handler or expert there to supervise. Ostriches, with their big eyes, long feathery eyelashes and fluffy soft feathers, can look very friendly and inviting. They are not aggressive birds, unless they are protecting their young chicks, but an ostrich is a big, strong, tall and wild bird. With their size and weight, and their curious nature, they can bump, step on or poke you with their strong beaks if you're not careful.

Meeting an ostrich in an ostrich farm

But if you go to such a place where there are ostriches, and an ostrich handler or zoo keeper gives their permission and supervises you, meeting an ostrich up close is an experience you'll never forget.

ABOUT THE AUTHOR

Novare Lawrence loves researching and writing books about Nature. She shares the knowledge and beauty of our natural world with kids young and old hoping that we will all do our part to help preserve our planet and all the wonderful species upon it.

You may learn more about her books and the *A Bird Book for Kids*™ series at her website:
ABirdBookforKids.com
And at
NadaBinduPublishing.com

A Bird Book for Kids™ Books
by Novare Lawrence

Digital:
- Hummingbird
- Ostrich
- Falcon
- Owl
- Condor
- Peacock
- Quail

Print:
- Hummingbird and Ostrich

www.ingramcontent.com/pod-product-compliance
Lightning Source LLC
Chambersburg PA
CBHW060817270326
41930CB00002B/65